Book 1
CompTIA A+
By Solis Tech

&

Book 2
Robotics
By Kenneth Fraser

Book 1
CompTIA A+
By Solis Tech

All-in-One Certification Exam Guide for Beginners!

CompTIA A+: All-in-One Certification Exam Guide for Beginners!

Table Of Contents

Introduction......5

Chapter 1: The CompTIA A+ Examination......6

Chapter 2: The Different Parts of a Computer......8

Chapter 3: How to Work With Computer Parts Effectively......15

Chapter 4: The Tools Needed for Checking Computer Parts......19

Chapter 5: Operating Systems......22

Conclusion......25

Introduction

I want to thank you and congratulate you for purchasing the book, *"CompTIA A+: All-in-One Certification Exam Guide for Beginners!"*

This book contains proven steps and strategies on how to prepare for the CompTIA A+ exams.

This eBook will explain the basics of the CompTIA A+ certification and tests. It will also give you some pointers regarding the topics that you need to review. By reading this book, you will gain the knowledge and skills required to pass the tests.

Thanks again for purchasing this book, I hope you enjoy it!

Chapter 1: The CompTIA A+ Examination

This eBook is written for people who are knowledgeable about computers. It assumes that you know to how to use a computer and its peripherals (e.g. printers, modems, etc.). This book will serve as your guide in preparing for the CompTIA A+ exam.

The A+ Certification

This is a certification program developed by CompTIA (Computer Technology Industry Association). This program is designed to provide a consistent way of checking the competency of computer technicians. The A+ certificate is given to people who have reached the degree of knowledge and diagnostic skills required to give proper support in the PC industry.

The A+ certification is similar to other programs in the industry (e.g. Microsoft Certified Systems Engineer and Novell's Certified Novel Engineer). The principle behind these certification programs is that if you need to get services for their products, you want to find technicians who have been certified by these programs.

The Benefits of Being A+ Certified

There are many reasons to get your own A+ certification. The information packet distributed by CompTIA gives the following benefits:

- It serves as a proof of your professional achievement.
- It improves your marketability.
- It gives you excellent advancement opportunities.
- It is now considered as a requirement for other kinds of advanced computer training.
- It encourages customers to do business with you

How to Become Certified

The A+ certification is given to anyone who passes the exams. You are not required to work for any company. CompTIA is not a secret group or society. It is, on the other hand, a group of elite computer technicians. If you want to be A+ certified, you have to do these things:

- Pass the exam called A+ Essentials
- Pass one of the three technician examinations:

- IT Technician Test
- Depot Technician Test
- Remote Support Technician Test

You can take the tests at any Pearson VUE or Thompson Prometric testing center. If you will pass both exams, you will receive a mail from CompTIA. That letter will inform you that you passed the tests. Additionally, it contains the certificate, a lapel pin, and a business card.

How to Sign Up for the Exams

To sign up for the tests, you may call Pearson VUE at 1-877-551-7587 or register online at www.vue.com. For Thompson Prometric, call 1-800-777-4276 or visit the website www.2test.com.

These companies will ask for your name, employer, phone number, mailing address, and SSN (Social Security Number). If you don't want to give out your SSN, a provisional number will be given to you. Additionally, they will ask when and where you want to take the exam.

Obviously, the exams aren't free. You have to pay your chosen testing company. That means you have to specify the payment arrangement during the registration process. You can simply provide your credit card information to the customer representative you will talk to. If you're doing it online, you can enter the credit card info on their payment page.

Who Should Use This Book?

If you want to pass the A+ tests, and do it confidently, you should use this book as a guide for your preparations. The A+ Essentials test is created to measure basic skills for an entry-level computer technician. The technician tests are designed to certify that you have the required skills to service microcomputer hardware.

This eBook was created with one purpose in mind: to help you pass the A+ exams. This guide will do that by explaining the things on which you will be tested.

Chapter 2: The Different Parts of a Computer

A PC (i.e. personal computer) is a machine made up of different components that work together to perform tasks (e.g. helping you write a document or add up large numbers). With this definition, notice that computers are described as having various distinct parts that work together harmoniously. Nowadays, almost all computers are modular. That is, they possess parts that can be replaced if the owner wants to improve the performance of his device. Each part has a specific purpose. In this chapter, you'll learn about the parts that make up a common PC, how they work, and what their functions are.

Important Note: Unless stated otherwise, the terms "computer" and "PC" can be used interchangeably throughout this eBook.

The Different Parts of a Motherboard

The motherboard, also called the system or planar board, serves as the "spine" of a PC. This is the brown or green circuit board that you'll find at the bottom of your computer. The system board is the most important part of a PC since it houses and/or connects the other parts of a computer together.

Different Types of Motherboards

There are two main types of motherboards. These are:

Integrated Motherboards – With this type, most of the parts are integrated into the system board's circuitry. Basically, integrated motherboards are created for simplicity. Since majority of the components are already part of the board itself, you won't have to install them individually. However, this simplicity has a major drawback: once a component stops working, you cannot simply replace it; you have to replace the entire motherboard. These boards are cheap to manufacture but expensive to repair.

Note: If one of the parts breaks, you may just disable it and add an expansion card that has similar capabilities.

Nonintegrated Motherboards – Here, the major parts (e.g. disk controllers, video circuitry, etc.) are installed as expansion cards. You will easily identify this kind of system board since every expansion slot is occupied by a major component.

The Different Form Factors of Motherboards

Computer experts also classify motherboards according to their design (also known as *form factor*). Here are the main form factors being used today: NLX, BTX, ATX, and micro ATX. You have to be vigilant when buying a computer case and system board separately. Some cases lack flexibility: they might not accommodate the system board you will select.

Let's discuss each form factor:

1. NLX – This is the abbreviation for "New Low-profile Extended". In general, this form factor is used for cases that are low-profile. With this design, the expansion slots (e.g. PCI, ISA, etc.) are placed on a special card to reduce the vertical space they occupy. Daughter boards, or adapter cards, that are normally plugged vertically into the expansion slots, are placed parallel to the system board. That means their size won't affect that of the computer case.

2. BTX – This form factor was launched by Intel back in 2003. With this design, the head-producing parts are lined up against the power supply's exhaust fan and the air intake vents. Then, the other components are cooled by installing heat sinks on the motherboard. This design offers a quiet setup since it involves efficient airflow paths and fewer exhaust fans.

3. ATX – With ATX motherboards, the processor and memory slots form a 90° angle with the expansion cards. This design places the memory and processor in line with the power supply's exhaust fan. Thus, the processor can remain cool while it runs. In addition, you may add expansion cards (even the full-length ones) to an ATX motherboard since the memory and processor are not parallel to the expansion cards.

4. Micro ATX – This form factor is similar with the previous one, with one major difference: it is designed for smaller computer cases. Micro ATX motherboards benefit from the enhanced cooling designs of their full-sized counterparts. However, since they are smaller, they have lesser motherboard headers, integrated components, expansion slots, and memory modules.

Processors – Their Functions and Characteristics

Now that you are familiar with system boards, you have to learn about their most important part: the central processing unit (CPU). The CPU controls all of the computer's activities using both internal and external buses. Basically, it is a processor chip that contains millions of transistors.

Important Note: Nowadays, the word "chip" describes the whole package that a computer technician may install into a socket. However, this word was originally used to refer to the silicon wafer hidden inside the carrier (i.e. the "chip" you see on your motherboard). The pins that you see on the outer part of the carrier are connected to the silicon wafer's small contacts. These pins allow you to install the carrier into a socket.

You can identify which part inside the PC is the central processing unit: the CPU is a large square that lies flat on the motherboard with a large fan and heat sink.

The Features of Modern Processors

- Hyperthreading – This word refers to HTT (hyper-threading technology). Basically, HTT is a variant of SMT (simultaneous multithreading). This kind of technology uses the scalar architecture of modern CPUs.

  HTT-capable CPUs appear as two different processors to the computer's operating system (OS). Because of this, the OS may assign two processes simultaneously, such as symmetric multi-processing, where multiple processors utilize the same network resources. Actually, the OS should support SMP in order to use HTT. If a process fails because of missing information caused by, for instance, branch prediction problems, the processor's execution resources can be reassigned for a different procedure that can be conducted immediately. Thus, the processor's downtime is dramatically reduced.

- Multicore – A CPU that has a multicore design contains two processors inside the same package. Here, the OS may treat the CPU as if it were two different CPUs. Just like the HTT, the OS should support SMP. Additionally, SMP is not considered as an upgrade if the apps run on the SMP system are not meant for parallel processes. A good example for the multicore technology is the i7 Quad-Core Processor from Intel.

- Microcode – This is the group of instructions (also called instruction set) that compose the different microprograms that the CPU executes as it performs its functions. The MMX (multimedia extensions) is a special example of an individual microprogram that performs a specific instruction set. Basically, microcodes are at a lower level than the codes used in computer programs. On average, each instruction from a computer program requires a large number of microinstructions. Intel and other processor manufacturers incorporate the MMX instruction set into their products.

- Overclocking – This feature allows you to increase the performance of your CPU, on par with processors created to function at overclocked rates. However, unlike processors created to function on that speed, you have to make sure that the overclocked processor doesn't damage itself from the increased level of heat. You might need to install an advanced cooling system (e.g. liquid cooling) to protect the CPU and other computer parts.

- Throttling – Processor throttling, also called clamping, is the process that specifies the CPU time to be spent on a computer program. By specifying how individual programs use the processor, you can "treat" all of the applications fairly. The principle of Application Fairness turns into a major problem for servers, where each program may represent the work of another user. That means fairness to computer programs becomes fairness to the users (i.e. the actual customers). Customers of modern terminal servers take advantage of this feature.

Memory – Its Functions and Characteristics

Nowadays, memory is one of the easy, popular, and inexpensive methods to enhance a computer. While the computer's processor runs, it stores data in the machine's memory. Basically, the more memory a machine has, the faster it can operate.

To determine the memory of a computer, search for thin sets of small circuit boards that are packed together near the CPU. These circuit boards sit vertically on the computer's motherboard.

How to Check for Errors in a Computer's Memory

Parity Checking

This is a basic scheme used to check for errors. It lines up the computer chips in a single column and separates them into equal bit groups. These bits are numbered beginning at zero. All of the number *x* bits, one from every chip, create a numerical array. If you are using "even parity", for instance, you will count up the number of bits contained in the array. If the total number is even, you will set the parity bit to zero since the bit count is already even. If the total is an odd number, on the other hand, you should set the parity bit to 1 in order to even up the bit count.

This technique is effective in identifying the existence of errors in the arrays of bits. However, it cannot indicate the location of the errors and how to solve them. Keep in mind that this isn't error correction – it is just a simple error check.

ECC

ECC stands for *Error Checking and Correcting*. If the computer's memory supports this method, the system will generate and store check bits. Whenever the machine accesses its memory, an algorithm will be performed on the check bits. If the result turns out to be zero (or a group of zeros), the information contained in the memory is considered valid and the computer functions as normal. ECC can identify single-bit and double-bit errors. However, it can only correct errors that are single-bit in nature.

The Four Main Types of Memory

- DRAM – This is perhaps the most popular type of RAM out there. DRAM stands for *Dynamic Random Access Memory*. Because of their inherent simplicity, these memory chips are cheap and easy to create compared to the other types. This kind of memory is called dynamic since it needs constant update signals in order to keep storing the data written there. If the DRAM chips won't receive stable signals, the information they hold will be deleted.

- SRAM – This stands for *Static Random Access Memory*. Unlike DRAMs, this kind of memory doesn't require a steady stream of signals. In general, SRAM chips are more complex and expensive than DRAMs. You can use SRAM for cache functions.

- ROM – This is the abbreviation for Read-Only Memory. It is called as such because it prevents the user from editing the memory it contains. Once the data is written on the computer's ROM, it cannot be changed anymore. ROM is usually used to hold the machine's BIOS, since this data is rarely modified.

- CMOS – This is a special type of memory chip. It is designed to hold the configuration settings of a computer's BIOS. CMOS is battery-powered: that means the configuration is retained even if the machine is turned off.

Storage Devices – Their Functions and Characteristics

Computers are useless if they can't store anything. Storage devices hold the information being used, as well as the programs and files the computer needs in order to function properly. In general, storage devices are classified according to their capacity, access time, and physical attributes.

HDD Systems

HDD stands for *Hard Disk Drive*. This storage device is also called hard disk or hard drive. Computers use HDDs to allow quick access to data as well as permanent storage. Typically, hard disks are found inside a computer.

An HDD system is composed of:

Controller – This component controls the storage. It knows how the drive functions, emits signals to the different motors inside the disk, and accepts signals from the sensors within the drive. Nowadays, hard disk manufacturers place the drive and controller in one enclosure.

Hard Disk – This acts as the physical warehouse for the data. HDD systems store data on little disks (about 3-5 inches in diameter) grouped together and kept inside an enclosure.

Host Adapter – This is the system's translator: it converts signals from the controller and hard disk to signals the computer can work with. Most modern motherboards have a built-in host adapter, allowing drive cable connection through board headers.

Floppy Drives

Floppy disks are magnetic storage devices that use plastic diskettes enclosed in a tough casing. Several years ago, floppy disks were used to easily transfer information from one computer to another. Nowadays, few people are using floppy disks because of their small capacity. DVD-ROMs and CD-ROMs have replaced floppy disks in storing and transferring digital information.

CD-ROM Drives

Modern computers use CD-ROM drives. These compact disks are virtually similar to those used in music recording. CD-ROMs allow you to store data for a long period of time. In general, these drives are read-only: you cannot erase or delete the data once it is stored on a CD. In addition, computers need to spend a longer time in "reading" CDs compared to internal hard drives. Why are these drives so popular?

Despite their drawbacks, CD-ROMs are used because they can store large files (about 650MB) and are extremely portable.

DVD-ROM Drives

This is the newest storage device to be used for computers. The DVD (i.e. digital video disc) technology is mostly used for entertainment purposes (e.g. home theater systems). DVD-ROMs are basically similar to the DVDs you use at home. Because of this, computers that are equipped with a DVD-ROM drive can play movies stored on a DVD.

However, DVD-ROMs are way much more useful when used for computers. Since they use newer technology, DVD-ROMs are better than CD-ROMs in terms of storage capacity. On average, DVDs can hold 4GB of data. That means DVD-ROMs are your best option if you are storing or distributing large files.

Important Note: CD-ROMs and DVD-ROMs have the same appearance. The single difference is the logo on the front of DVD drives.

Removable Storage Devices

Many years ago, the term "removable storage" meant something extremely different from what it means now. Tape backup is one of the old storage devices that can still be bought today. Modern computer users prefer the solid-state, random-access removable storage devices. In this section, you'll learn about tape backups and the new storage solutions.

Tape Backup

This is an old type of removable storage. A tape backup device can be installed externally or internally and utilize either an analog or digital magnetic tape to store data. In general, this kind of device can hold more information than other storage mediums. However, they are also one of the slowest in terms of data transfer rate. Because of these reasons, tape backup devices are mainly used for archived information.

Flash Memory

Before, random-access memory chips were only used to access and use data. But now, you'll find them in different physical sizes and storage capacities. Flash memory drives are considered as the best solid-state storage device available. The flash memory category includes SD (secure digital) and other memory cards, USB flash drives, and older detachable and non-detachable memory mechanisms. Each of these storage devices has the capability to store huge amounts of information.

Manufacturers of flash memory devices use revolutionary packaging (e.g. keychain attachments) for their products to provide easy transport options for their end-users.

Chapter 3: How to Work With Computer Parts Effectively

While taking the CompTIA A+ exam, you will answer questions regarding the installation, usage, and replacement of computer parts. This chapter will help you to review regarding those topics.

How to Install, Configure and Optimize Computer Parts

Aside from knowing the characteristics and functions of PC components, you also need to know how to use them. In particular, you should be familiar with the installation, configuration, and optimization of such parts.

How to Upgrade a Storage Device

Storage devices are available in different shapes and sizes. Aside from IDE and SCSI, two of the most popular types, there are SATA (Serial ATA) and PATA (Parallel ATA). You can also differentiate between external and internal drives. This section of the book will explain each of these options.

Preparing the Drive

Regardless of the technology being used, you should format storage devices before using them. Although most drives have their own formatting software, each OS has a tool that you can use. When working with Windows computers, you can utilize the format utility through the command line. If you are working with XP, Vista, 7, or newer Windows system, you can also use the graphical utility program called Disk Management.

How to Work with IDE

Before, IDE (integrated drive electronics) drives were the most popular kind of computer hard drives. Although they are often linked to hard drives, IDE is more than just an interface for hard disks. It can also serve as the interface for different storage types such as Zip, DVD, and CD-ROM.

To install IDE drives, you should:

1. Set the slave/master jumper on the IDE drive.

2. Place the drive inside the drive bay.

3. Connect the cable for power-supply.

4. Link the ribbon cable to the motherboard and to the drive.

5. If the drive isn't detected automatically, you should configure it using the BIOS Setup of your computer.

6. Use your PC's operating system to format and partition the IDE drive.

How to Work with SCSI

SCSI is the abbreviation for *Small Computer System Interface*. This kind of device can be either external or internal to the machine. To configure an SCSI device, you should assign an SCSI ID (also called SCSI address) to all of the devices in the SCSI bus. You can configure their numbers using a jumper or DIP switch.

Whenever the computer sends data to the SCSI device, it emits a signal on the cable assigned to that number. The device will respond with a signal that holds the device's number and the information needed.

You should install a terminator (i.e. terminating resistor device) at the two ends of the bus to keep the SCSI devices working. You can activate and/or deactivate terminators using a jumper.

Here are the things you should do when installing an SCSI device:

- For Internal Devices – Connect the cable (i.e. a 50-wire ribbon cable with multiple keyed connectors) to the adapter and to each SCSI device in your computer. Afterward, place the terminators on the adapter and terminate the final device in the chain. You should leave other devices unterminated.

- For External Devices - Follow the steps outlined above, but here, you should use some stub cables between the SCSI devices in the daisy chain (instead of a long cable that has multiple connectors). Terminate the adapter as well as the final device in the daisy chain (that device should have one stub cable linked to it).

- For Hybrid Devices – Many types of adapters have connectors for external and internal SCSI devices. If you have this kind of adapter, attach the ribbon cable to your internal devices and connect the cable to your adapter. Afterward, daisy-chain the external devices from the external port. Terminate the device at the end of every chain. Make sure that the adapter is unterminated.

External Storage Devices

As capacities shoot up and prices fall down, the number of available external storage devices has increased greatly. Aside from the SCSI variant explained above, you will also see devices with USB connections and those that can connect straight to the system. The computer's operating system will recognize USB devices upon connection. You can just install any additional programs you like to use. A computer program called Dantz Retrospect is included in many storage devices to allow you to utilize external devices as automatic backups.

If the external storage device is linked straight to the system, you can just follow the instructions that came with that product. Then, install additional programs on the computers that you will be using. The main benefit of linking straight to the system is that the storage device/s can be accessed by all of the computers.

How to Upgrade Display Devices

Before linking or unlinking a display device (e.g. a computer monitor), make sure that the computer and the device itself are powered off. Afterward, connect a cable from the computer's video card to the display device. Connect the power cord of that device to an electrical outlet. You may use a modern Digital Visual Interface (DVI) cable or the traditional DB-15 (or VGA) cable.

While installing a new monitor, you should have the proper driver. The driver is the software interface between the display device and the computer's OS. If you don't have the right driver, your monitor won't display what you want to see. Nowadays, you can download the newest drivers from the website of monitor manufacturers.

Aside from the power supply, the most dangerous part to repair is the monitor. Computer technicians say that beginners should never attempt to repair monitors. Monitors can hold high-voltage charges even if they have been powered off for several hours. That means you can be electrocuted if you will try to repair a monitor by yourself. If your monitor stopped working, and you don't want to buy a new one, you should take that device to a TV repair shop or a certified computer technician. The technicians and the repair guys know how to fix monitors properly – they understand the dangers and the correct procedures.

How to upgrade Input and Multimedia Devices

The typical upgrade for input devices is the transition to newer mice and keyboards.

Keyboards

Keyboards may wear out if used repeatedly. The usual problem is "key sticking", where keys are no longer responding to the user. To replace a PS/2 101-key keyboard with a new one, just unplug the old keyboard and plug in the new. As you can see, this is a quick and easy process. Nowadays, however, computer users prefer to replace old keyboards with USB ones.

Here is a principle you need to remember: You can use the "unplug-the-old-and-plug-in-the-new" procedure as long as your computer's OS supports the keyboard you want to use.

Mice

Computer mice also wear out because of repeated use. But don't worry: you can replace old mice with new ones. You may easily replace a PS/2 connection mouse with another without spending too much. As an alternative, you may buy an optical mouse (which prevents dust- and ball-related problems) or a wireless one (which needs batteries to send and receive signals). Although new mouse models still use the PS/2 type of connection, most mouse products in the market use the USB connection.

Chapter 4: The Tools Needed for Checking Computer Parts

The CompTIA A+ exam will also test your skills in checking computer parts. This chapter will help you with that topic by discussing the tools and diagnostic procedures needed.

The Tools Needed by a Computer Technician

A great computer technician needs a great collection of tools. If you are working alone, you may not get past the troubleshooting phase. However, you still need to use certain tools in order to succeed in that task. Once you have identified the problem, you will need to get another set of tools in order to fix it.

This book will focus on the "hardware" tools. These are:

- Screwdrivers – When checking a computer technician's toolkit, you will surely find screwdrivers. Almost all of the big computer parts you'll see today are mounted using screws. If you need to remove these parts, you need to have the right type of screwdriver. This kind of tool is divided into three types:
    - Flat-Blade – Many people refer to this as the *common* or *standard* screwdriver. The screw used with this screwdriver is rarely used today (mainly because the screw's head can be destroyed easily).
    - Phillips – This is perhaps the most popular type of screwdrivers being used today. The screws used with a Phillips screwdriver have enough head surface: you can turn them many times without damaging the screws' head. According to recent reports, more than 90% of the screws used in computers belong to the Phillips-head type.
    - Torx – This is the type of screwdriver you use while working on tiny screws found on Apple and Compaq computers. The screws you remove using a Torx screwdriver have the most surface to work against: they offer the best resistance in terms of screw-head damage. Nowadays, Torx-type screws are gaining more popularity because of their clean and technical look.
- Flashlight – This is one of the tools you should always have. You'll realize how important this tool is when you're crawling under a table searching for a dropped computer part.

- Needle-Nose Pliers – You should have this in your toolkit. This kind of pliers is great for holding connectors or tiny screws (particularly if you have large hands). If needle-nose pliers are still too big to do certain tasks, you may use a pair of tweezers.

- Compressed Air – While working on a computer, you will usually remove the machine's case first. Once the cover is removed, it would be great if you will clean the computer's internal components. The clumps of dirt and fibers can block airflow inside the system unit. As a result, the PC's life will be shortened. The ideal way to eliminate the dust is by using compressed air.

  If you are working for a big company, you probably have a core air compressor that supplies compressed air. If this kind of compressor is not available, you may purchase canned compressed air. However, you'll be shelling out large amounts of money – cans of compressed air are expensive.

- Soldering Iron – You can use it to splice broken wires. Nowadays, computer technicians rarely use this tool. Here's the reason: modern computer parts are created with quick-disconnect connectors. You can easily replace them without splicing anything.

- Wire Strippers – Whenever you have to solder something, you need to use a stripper/wire cutter to prepare the wires for connection. Stripping means you will expose a certain part of the wire by removing the insulation.

- Multi-Meters – This tool is named as such because it is basically a set of different types of testing meters, such as ammeter, voltmeter, and ohmmeter. When used by a trained technician, a multi-meter can identify the failure of various types of computer parts.

  A multi-meter has an analog or digital display, a mode selector switch, and two probes. You can use the switch to perform two things: (1) select the function you want to test and (2) choose the range in which the meter will work. If you need to use an old meter to measure a power pack, you should manually set the range selector. Modern multi-meters, particularly the digital ones, can automatically find the correct ranges.

  Important Note: You should never measure voltage by connecting a manual ranging multi-meter to an AC electrical outlet. This will damage the meter itself, the meter mechanisms, or both.

<u>How to Measure Resistance Using a Multi-Meter</u>

Resistance is the property of electricity commonly measured when troubleshooting computer parts. This electrical property is measured in ohms and represented by the Greek letter "omega." If a multi-meter indicates infinite resistance, the electric currents cannot travel from one prove to another. If you are using a multi-meter to check the resistance and you are getting an infinite reading, there's a huge possibility that the wire is broken.

When measuring resistance, you should set the tool to measure ohms. You can do it using either the selector dial or the front button. Then, connect the PC component you want to measure between the tool's probes. The multi-meter will then show the component's resistance value.

How to Measure Voltage Using a Multi-Meter

This process is similar to the one discussed above, but with two main differences:

1. While measuring voltage, make sure that you properly connect each probe to the source of electricity. For DC voltage, the "-" should be connected to the negative side and the "+" to the positive one. This positioning is irrelevant when measuring AC voltage.

2. You should switch the selector to Volts DC (VDC) or Volts AC (VAC), depending on what you need to measure, to instruct the tool about the voltage you are working with. These settings protect the tool from overload. The multi-meter will blow up if you will plug it into a power source while it's still on "measure resistance" mode.

Chapter 5: Operating Systems

The CompTIA A+ examination will test your knowledge regarding operating systems. Since operating systems play an important role in the computer industry, you should be familiar with them. This chapter will guide you in this topic. Here, you'll learn different things about a computer's OS.

What is an Operating System?

Computers are useless if they don't have any piece of software. Well, you can use them as a doorstop or paperweight – but that is not cost-efficient. You need to have an interface before you can use the capabilities of a computer. And, if you don't know yet, software acts as the interface. Although there are different kinds of software, or computer programs, the most important one you'll ever need is the OS.

Operating systems have various functions, most of which are extremely complex. However, two functions are critical:

1. Interfacing with the computer's hardware

2. Providing an environment in which other pieces of software can run.

Here are the three main types of software that you will encounter in the CompTIA exam:

- Operating System – It provides a stable environment for other computer programs. In addition, it allows the user to enter and execute commands. The operating system gives the user an interface so they can enter commands (i.e. input) and get results or feedback (i.e. output). For this, the OS should communicate with the PC's hardware and conduct the tasks below:

  - Device access

  - Output format

  - Memory management

  - File and disk management

  Once the operating system has performed these basic tasks, the user can enter instructions to the computer using an input device (e.g. a mouse or keyboard). Some of the commands are pre-installed in the operating system, whereas others are given using certain applications. The OS serves as the platform on which the PC's hardware, other pieces of software, and the user work together.

- Application – This is used to complete a specific task. Basically, an application is a computer program written to support the commands given to the OS. Every application is compiled or configured for the operating system it will be used for. Because of this, the application depends on the OS to perform most of its basic functions.

  When a program accesses the computer's memory and linked devices, it sends a request to the OS. The machine's operating system will perform the requests made by the program being used. This setup helps greatly in decreasing the programming overhead, since most of the executable codes are shared – they are written onto the operating system and can be used by different applications installed on the computer.

- Driver – This is an extremely specific program created to instruct an operating system on how to access and use a piece of hardware (e.g. webcam, flash memory, etc.). Every webcam or flash memory has distinct features and settings – the driver helps the OS in knowing how the new hardware works and the things it can do.

The Terms and Concepts Related to Operating Systems

In this section, let's define some of the most important terms and concepts. Study this section carefully since it will teach you the terms you'll encounter during the CompTIA A+ exam.

<u>*Key Terms*</u>

- Source – This is the code that explains how computer programs work. An operating system can be open source or closed source.
    - Open Source – The users have the right to change and examine the code.
    - Closed Source – The users are not allowed to edit or check the code.
- Version – This is a specific variant of a computer program, usually expressed by a number, which informs users regarding the "newness" of the software. For instance, MS-DOS is now in its sixth main version. Computer programmers distinguish minor revisions from major ones this way:
    - "Program" 4.0 to "Program" 5.0 is a major revision.
    - "Program" 5.0 to "Program" 5.2 is a minor revision.

- Shell – A piece of software that works on top of the operating system. It allows users to execute commands through an array of menus or a different type of graphical interface. A shell makes an operating system simpler and easier to use by modifying the GUI (graphical user interface).
- GUI – The method by which a user communicates with computers. A GUI uses a touchpad, mouse, or a different mechanism (aside from a keyboard) to interact with the machine and issue commands.
- Multithreading – The capability of a computer program to contain several requests in the computer's CPU. Since it allows an application to perform different tasks simultaneously, computers experience a boost in performance and efficiency.
- Network – A group of computers that are connected by a communication link. A network allows computers to share resources and information.
- Preemptive Multitasking – This is a multitasking technique in which the operating system allocates each program a certain amount of CPU time. Afterward, the OS takes back the control and provides another task or program access to the CPU. Basically, if a computer program crashes, the operating system takes the processor from the faulty program and gives it to the next one (which must be working). Even though unstable computer programs still get locked, only the affected application will stop – not the whole machine.
- Cooperative Multitasking – This is a multitasking technique that relies on the applications themselves. Here, each program is responsible for utilizing and giving up access to the CPU. Windows 3.1 used this method to manage multiple programs. If an application stalls while it is using the CPU, the application fails to free the CPU properly, and the whole computer gets locked, the user needs to reboot the machine.

Conclusion

Thank you again for purchasing this book!

I hope this book was able to help you to prepare for the CompTIA A+ tests.

The next step is to reread this book and use other information sources. That way, you can increase your chances of passing the exam.

Finally, if you enjoyed this book, please take the time to share your thoughts and post a review on Amazon. It'd be greatly appreciated!

Thank you and good luck!

Book 2
Robotics
By Kenneth Fraser

The Beginner's Guide to Robotic Building, Technology, Mechanics, and Processes

Robotics: The Beginner's Guide to Robotic Building, Technology, Mechanics, and Processes

Table of Contents

Introduction....29

Chapter 1 - What is a Robot?....30

Chapter 2 - Getting Started....34

Chapter 3 - Understanding Actuators....39

Chapter 4 - Microcontrollers and Motor Controllers....43

Chapter 5 - Controlling Your Robot and Use of Sensors....46

Chapter 6 - Assembling and Programming a Robot....49

Conclusion....54

Introduction

I want to thank you and congratulate you for purchasing the book, *"Robotics: The Beginner's Guide to Robotic Building"*.

This book contains proven steps and strategies on how to build your own robot that will perform certain functions as you want it to do.

For most people, a robot is a machine that could mimic a human such as R2D2 and C3PO in Star Wars. But these types of robots are still in the figments of our imaginations. We are still far from giving robots high level of artificial intelligence to easily adapt and interact to its environment. There is however pioneering works on artificial intelligence that hopes to create humanoid robots.

The type of robots that exist and working today are robots that are programmed to do things that are too dangerous for humans, too repetitive, or just plain messy. These robots are often found in wide range of industries and places such as oil refineries, hospitals, laboratories, factories, and even in the Outer Space. There are about more than a million robots are working in different fields today.

There are types of robots that bring joy to kids such as the popular AIBO ERS-220 that is a bestseller toy during Christmas. While some robots perform great feats by discovering new places and gathering important data in the name of science, specialized robots such as the Mars Rover Sojourner and the underwater robot Caribou are sent to places that average humans cannot go.

Robots are exciting machines to play with, but they are more exciting to build. For hobbyists, building their own robots that capable of doing whatever they program these machines to do gives them pure delight.

This book introduces you to the science of robotics – its basic elements and fundamental concepts. And at the course of your reading, you will learn all the essential aspects you need to build your own robot.

Thanks again for purchasing this book, I hope you enjoy it!

Chapter 1 – What is a Robot?

It is interesting to know that even with all the hype about robots and with all the milestones in robotics, there is still not standard definition for a robot. There are, however, some basic characteristics that a robot should have and this could help you determine if a certain object is a robot or not. It will also guide you in deciding what features you need to build into a machine before you can say that it is a robot.

Four Basic Characteristics of a Robot

A robot has four basic characteristics: sensing, movement, intelligence, and energy.

Sensing

Basically, a robot must be capable of sensing its environment much similar to the way humans sense its surroundings. Robots could either sense through light sensors that mimic the functions of the eyes, or be equipped with chemical sensors that function like the nose, sonar sensors like the nose, touch sensors like the skin, and taste sensors like the tongue. These sensors will help the robot to become aware and understand its environment.

Movement

A robot should have the ability to move around through walking on legs, rolling on wheels, or through propellers. It's either the entire robot is able to move or just some parts of it such as head, arms, or just legs.

Intelligence

A robot should be equipped with artificial intelligence or AI. This is usually done through computer programing. Hence, you need a background in programming to provide your robot with the needed intelligence. You need to program the robot's intelligence so that it will know what to sense and how to move.

Energy

A robot should have a way to power itself. The energy source could be electrical, chemical (battery), or solar. The method by which your robot energizes itself depends on what your robot is required to do.

Working Definition of a Robot

For the purpose of discussion and for reference, we define robot as a machine that contains control systems, sensors, manipulators, software and power supplies that works together to do certain tasks.

Building a robot requires understanding of the fundamental principles of mechanical engineering, mathematics, physics, and computer programming. In special cases, it also requires specific knowledge on chemistry, biology, and medicine. In studying robotics, you need to be actively engaged with wide range of disciplines to build robots that could solve certain problems.

A Brief History of Robotics

The word "robot" was first used in a play entitled R.U.R (Rossum's Universal Robots) written in 1921 by Czech writer Karl Capek. This play is about machines that are built to work on a factory and eventually revolted against their human masters. Robots are the Czech word for slave.

Meanwhile, the word robotics also first appeared in a work of fiction. Russian-born American fictionist Isaac Asimov used it in his short story "Runabout" (1942). Compared to Capek, Asimov had a more positive opinion of the role of robots in the society. In general, he described robots as useful machines that serve humans and perceived them as a "better, cleaner race. He also proposed the three Laws of Robotics:

First Law of Robotics

A robot may not injure a human or, through inaction, allow a human to come to harm.

Second Law of Robotics

A robot must obey the orders given by humans except if such orders would violate the First Law.

Third Law of Robotics

A robot may protect itself as long as such protection do not violate with the First Law or Second Law of Robotics.

Early Models of Robots

Among the earliest cases of a mechanical system designed to perform a regular task was recorded around 3000 BCE. Egyptian water clocks are added with human statuettes to hit the hour bells and signal the passing of time. In 400 BCE,

Archytus of Taremtum, who was known as the inventor of pulley and screw, created a pigeon made of wood that is capable of flying. Meanwhile, hydraulically-powered figurines that could speak prophecies were common during the Greek domination of Egypt during the second century BCE.

In the first century C.E., Petronius Arbiter built a doll that is capable of moving like a human being. In 1557, Giovanni Torriani built a wooden robot, which could fetch the Emperor's bread every day from the store. By 1700s, robotic inventions became common with numerous impractical yet ingenious machines such as steam-powered automata crafted in Canada as well as the popular talking doll by Thomas Edison. Even though these creations may have inspired the design and functions for the modern robot, the progress during the 20th century in the field of robotics exceeded previous advancements many times over.

The First Modern Robots

The robots that we are familiar with were built by George C. Devol in the 1950s. The inventor from Kentucky designed and patented a reprogrammable manipulator that he dubbed as "Unimate" derived from "Universal Automation." For years, he tried commercializing his product, but failed. But in 1960s, the entrepreneur-engineer Joseph Englberger bought the patent from Devol and modified it into an industrial robot. He established his company, Unimation, for production and marketing of these products. He was successful in this venture, and in fact, Englberger is regarded today as the Father of Robotics.

Robotics also progressed within the academic institutions. Alan Turing, pioneering computer scientist, mathematician, logician, and cryptologist, published his book "Computing Machiner and Intelligence" where he proposed a test to determine if a machine has the capacity to think for itself. This test is known as the Turing Test.

In 1958, Charles Rosen of the Stanford Research Institute created a research team to work in the development or a robot known as "Shakey" that was more advanced compared to Devol's Unimate. Shakey can move around through the room, sense light through his "eyes" move around strange environment, and to a particular degree, and react to what is happening to his surroundings. He was called Shakey because of his clattering and rickety motions.

In 1966 at Massachusetts Institute of Technology (MIT), Joseph Weizenbaum created an artificial program named ELIZA, which functions as a computer psychologist that manipulates its user's statements to formulate questions.

In 1967, Richard Greenblatt developed MacHack, a program that is capable of playing chess, as a response to a critical article written by Hubert Dreyfuss where he boasted that no computer program can beat him in chess. When the program is finished, Dreyfuss was invited to play and was defeated. This program was the foundation of future chess programs that eventually developed into Big Blue, the program that defeated Grand Master Gary Kasparov in 1997.

The interest in robotics is one of the major catalysts in the development of computers. In 1964, the IBM 360 becomes the first computer to be produced massively.

Robots are also crucial in pioneering space explorations. In 1969, the United States successfully used the latest technology in robotics and computing for Neil Armstrong's landing on the moon. Robots also helped in the expansion of scientific knowledge. In 1994, Carnegie Universities crafted Dante II, an eight-legged walking robot that successfully descends into the crater of Mt Spur to gather samples of volcanic gas.

Commercial companies also leveraged on the mass appeal of robots. In 1999, Sony released its original version of AIBO, a robotic dog that can entertain, learn, and communicate with its owner. Advanced versions have followed in the succeeding years, with the final model, the ERS-7M3, released in 2005.

Honda also released its ASIMO robot, an advanced humanoid robot in 2000. In 2004, Epsom was hailed as the world's smallest robot (7 cm high and weighs only 10 g.) The robotic helicopter is designed to fly and capture videos during natural disasters.

After being released in 2002, a robotic vacuum cleaner known as the Roomba became a huge hit. It sold more than 2.5 million units, which shows that there's really a huge demand for domestic robot technology.

Hundreds of films feature robots such as The Day the Earth Stood Still (1951), Arthur C. Clark's 2001: A Space Odyssey (1968), Star Wars (1977), Blade Runner (1982), Terminator (1984), Nemesis (1992), I, Robot (2004), Transformers (2007), and many more. The popularity of robotic films shows that people are inspired and delighted by the idea of machines that can independently move and think for itself.

If you are ready to build your own robot, continue to the next Chapter to help you get started.

Chapter 2 – Get Started

The first step in building your own robot is to determine what it should do, that is, your purpose of why you are building the robot. Robots can be used in different situations and are mainly designed to assist humans. It will help you a lot to learn first the different purposes and uses of robots.

Basically, robots are divided into two main groups: industrial and domestic robots.

Industrial Robots

Industrial robots are used in factories to manufacture products with precisions such as computers, cars, cellphones, medicine, and even food. Robots increased the productivity in different workplaces, which resulted to booming industries. Each type of industrial robot has its specific form that corresponds to its function. For instance, robotic arms are often used in car assembly lines to spray paint or weld frames. Robotic arms are among the most common robots today. Recently, agricultural robots have been introduced mainly to perform farm tasks such as cutting weeds and harvesting crops.

Domestic Robots

Domestic robots are mainly used in the home to perform household chores. They usually perform repetitive tasks every day such as vacuuming floors, mowing the lawn, vacuuming floors, and other chores that people usually don't have time to do. For example, there are vacuum robots that can clean the floors. They are equipped with motion sensors so they will not run into any object. You just need to push the switch on and it will do its job. It could pick up dust and pet hairs and could be used for hours.

There are also mower robots that could mow lawns. They are equipped with sensors to detect grass edges. Domestic robots are also used for entertainment such as Robosapien, AIBO, and iDog.

Choosing a Robotic Platform

The next step in building your own robot is to decide on the type of robot you want to build. A usual robot design usually begins with "inspiration" of what the robot will do and what it will look like.

The types of robots that you can build are endless. As long as you can envision something that a robot can do, you can work your way to achieve it. But for beginners, you can start with the following types: land robots, aerial robots, aquatic robots, stationary robots, and hybrid robots.

Land Robots

Land-based robots, particularly those added with wheels are among the most common mobile robots built by beginners, because they often require minimal investment while providing the opportunity to learn more about robotics. Meanwhile, the most advanced type of robot is the humanoid robot, which is akin to humans. Humanoids require several degrees of freedom and synchronization of different motors and use several sensors.

Wheeled Robots

Wheels are among the most common method of adding mobility to a robot and are used to mobilize many different sizes of robots and robotic platforms. Wheels could be about any size, and there's no limit in the number of wheels that you can add. More often than not, robots that are equipped with three wheels are using two wheels and a caster at one end. More advanced robots with two wheels are using gyroscopic stabilizing technology.

Meanwhile, robots that are added with four to six wheels usually use several drive motors that decreases the risk of slippage. Also, mecanum wheels or omni-directional wheels can provide the robot considerable benefits in mobility. Most beginners in robotic building are mistaken in thinking that inexpensive DC motors can mobilize robots that are medium in size. As you will learn later, there are more factors that you need to consider before you can add mobility to your robot.

Advantages

Wheeled robots are ideal for beginners as they are often more affordable to build. They have simple design and construction, and there are unlimited options. In addition, robots with six wheels or more could rival the mobility of a track system.

Disadvantages

Wheeled robots usually have small contact area, because only a small portion of the wheel is touching the ground. This results to lower traction that may cause slippage.

Tracked Robots

Tracks are used in tanks for mobility. Even though tracks, also known as treads, don't provide the added torque, they can decrease slippage and can equally distribute the robot's weight. This makes the robot easier to mobilize in loose ground such as gravel and sand. In addition, flexible track systems could easily navigate through a bumpy surface. Most hobbyists also believe that tank tracks are quite cool compared to wheels.

Advantages

Steady contact with the ground avoids slippage, which is prevalent with wheels. The track system also distributes weight evenly, which helps the robot in navigating different surfaces. Tracks can also be used to extensively enhance the ground clearance of the robot without adding a bigger drive wheel.

Disadvantages

The main disadvantage of using a track system for robots is that in turning, there's the tendency to cause damage to the surface that also causes damage to the tracks. In addition, robots are often built around the tracks, and there's a limit in the availability of the tracks. Drive sprocket can also considerably restrict the number of motors that you can use.

Legged Robots

More and more robots are using legs for movement. Legs are usually ideal to use for robots that should navigate on uneven ground. Many prototype robots are built with six legs that allow the robot for static balance. Robots with fewer legs are more difficult to balance as it requires dynamic stability. Once the robot ceases moving in the middle of the stride, it could fall over. Even though there were robots with one leg moving by hopping, bipeds, quadrupeds, and hexapods are the most common forms.

Advantages

The leg motion is the most natural among the platforms, and it can easily overcome big obstacles and move through rough surface.

Disadvantages

Most beginners are discouraged in building their first robot that moves using legs, as it requires high level of electronic, mechanical and coding skills. You also need to find a small battery that can provide the required power, so legged robots are usually expensive to build.

Aerial Robots

Humans have long been inspired by the idea of flight, and this transcends into the field of robotics. The idea of Autonomous Unmanned Aerial Vehicle (AUAV) has gained popularity over the years, and many enthusiasts have developed numerous prototypes. However, the benefits of crafting aerial robots have yet to prevail over the disadvantages. In building aerial robots, many hobbyists are still using commercial remote controllers. Professional aircrafts such as the Predator commissioned by the US military were partially autonomous though recently, updated versions of the Predator have completed aerial missions with only minimum human intervention.

Advantages

Aerial robots are great for surveillance, and remote controlled aircraft has been developed through the years, so there is a diverse community for mechanics where you can find support and know-how in building your own aerial robots.

Disadvantages

There is still limited community when it comes to autonomous control, as most of the knowledge on this field is protected by the US military. Meanwhile, this robot type is expensive as the whole robot could be broken if you miscalculate the steps and lead the robot into a crash.

Aquatic Robots

Recently, more and more hobbyists, communities, and companies are building unnamed aquatic vehicles. There are still many hindrances to overcome in order to make aquatic robots more enticing for the wider communities in robotics. But it is interesting to take note that there are companies today who are manufacturing robots that can clean pools. Aquatic robots can use thrusters, ballast, wings, tails, and fins to move under water.

Advantages

A massive part of the ocean is still unexplored so there's a lot to discover if you choose to build aquatic robots that could help in discovering the underwater world. The robot design is also guaranteed to be unique, and it could be tested in a pool.

Disadvantages

Aquatic robots are often very expensive to build, and there is the risk that the robot could be lost while deep in the ocean. You should also take note that most electronic parts don't pair well with water, especially salty water. You also need to consider the water pressure as going beyond deep sea needs considerable investment and research. There is also very limited robotic community that can provide support, and also limited wireless communication options.

Hybrid Robots

Your concept for the robot may not easily fall into any of the categories mentioned above or could be composed of various functional components. Take note that this book is written to guide you in building mobile robots and not those with fixed designs. In building a hybrid design, it is best to use a modular design where each functional component could be taken off and tested as a separate part.

Advantages

Hybrid robots are designed and built according to your preferences and needs. These robots could be used for various tasks and can be composed of modules. Hybrid robots could lead to versatility and increased functionality.

Disadvantages

Hybrid robots are often complicated to build and expensive. Parts need to be customized to fit the design.

Grippers and Arms

Even though grippers and arms don't fall under the category of mobile robots, robotics basically began with end-effectors and arms. Grippers and arms are the most ideal way for a robot to interact with the environment it is dealing with. Basic robotic arms could have just two to three motions; while more advanced arms could have more than a dozen movements.

Advantages

Most robotic arms and grippers have simple designs, and it is easy to make a three to four degree of freedom robotic arm with a turning base and two joints.

Disadvantages

Robotic arms are stationary unless you fix them on a mobile platform. The cost of building arms or grippers depends on the lifting capacity you need.

In the next chapter, you will learn how to choose the right actuators or motors for your robot.

Chapter 3 – Understanding Actuators

After learning general information about robots and robotics in the first two chapters, it is now time to choose the right actuators to mobilize your robot.

What Are Actuators?

Actuators are devices that transform energy into physical motion. In robotics, this energy is usually electrical energy. Most actuators today produce either linear or rotational motion. For example, a DC motor is a type of actuator.

Selecting the right actuator for your robot requires learning the available actuators, and some fundamental knowledge of physics and mathematics.

Rotational Actuators

Rotational actuators convert electrical energy into rotating motion. There are two primary mechanical parameters that distinguish each actuator: (a) the rotational speed that is often measured in revolutions per minute or rpm and (b) torque or the force that the devices can produce at a given distance often expressed in Oz•in or N•m.

AC Motor

Alternating Current (AC) is rarely used in robots because most of them are powered through Direct Current (DC) in form of cells or batteries. In addition, electronic parts use DC, so it is easier to use the same type of power supply for the actuators. AC motors are primary used in industrial settings where high torque is necessary or where the motors are connected to a wall outlet.

DC Motor

DC motors are often cylindrical in shape but they also come in different shapes and sizes. They also have output shafts that rotate at high speed often between 5000 and 10000 rpm. Even though DC motors rotate very fast, most have low torque. To decrease the speed and add torque, a gear could be added. To install a motor into a robot, you must fix the body of the motor to the robot's frame. Hence, motors usually have mounting holes that are basically located on the motor's face so that they can be easily installed. DC motors could either rotate in counter clockwise or clockwise. The angular movement of the turning shaft could be measured using potentiometers and encoders.

Geared DC Motor

A DC Motor could be added with a gearbox to reduce the motor's speed and enhance its torque. For instance, if a DC motor rotates at 5000 rpm and produces a 0.0005 N•m of torque, adding a 123:1 ("one hundred and twenty three to one") gear would reduce the speed by a factor of 123 (resulting to 5000 rpm / 123 = 40 rpm) and increase the torque by a factor of 123 (0.0005 x 123 = 0.0615 N•m). The most common types of gears are planetary, spur, and worm. Similar to a DC motor, a geared DC motor can also rotate in either clockwise or counter clockwise. You can add an encoder to the shaft if you want to know the number of rotations of the motor.

Hobby Servo Motors

Hobby Servo Motors, also known as R/C Servo Motors are actuators that rotate to a certain angular position, and were traditionally used in more expensive remote controlled machines for controlling or steering flight surfaces. Today, they are used in different applications so their prices have been reduced considerably, and the variety has also increased. Most servo motors can only rotate about 180 degrees. A hobby servo motor is composed of a DC motor, electronics, gears, and a potentiometer that measures the angle. The latter works with the electronics to mobilize the motor and stop the output shaft at a certain angle. In general, these servos have three wires, voltage in, control pulse, and ground. A robot servo is a recently developed servo that provides both position feedback and continuous rotation. Servos could rotate clockwise or counterclockwise.

Stepper Motors

As the name implies, stepper motors rotates following certain steps or degrees. The number of degrees the shaft rotates with every step could vary depending on various factors. Majority of stepper motors don't include gears, so similar to a DC motor, the torque is quote low. Fixing gears to a stepper motor has similar effect as installing gears to a DC motor.

Linear Actuators

Linear actuators produce linear movements. They have three primary distinctive mechanical properties: (a) the force measured in kg or lbs (b) speed measured in m/s or inch/s and (c) the maximum and minimum distance that the rod could move also known as the stroke measured in inches or mm.

Linear DC Actuator

A linear DC actuator is usually composed of a DC motor attached to a lead screw, which also turns as the motor moves. The lead screw has a traveler that is forced either away or towards the motor, basically transforming the rotating motion to a linear movement. Some DC linear actuators integrate a linear potentiometer that adds a linear position feedback.

Solenoids

Solenoids are comprised of a coil wound surrounding the mobile core. Once the coil is energized, the core is forced away from the magnetic field and creates a motion in one direction. Several coils or some mechanical arrangements will be needed to provide movements in different directions. A solenoid stroke is often very small but they are often very fast. The strength primarily depends on the size of the coil and the electrical power passing through it.

Hydraulic and Pneumatic Actuators

Hydraulic and pneumatic actuators use liquid or air respectively to create a linear movement. These actuators could have lengthy strokes, high speed and high force. To use these actuators, you need to use a fluid or air compressor that makes them harder to use compared to basic electrical actuators. These are often used in industrial applications because of their large size and high force speed.

Muscle Wire

Muscle wire is a specialized wire, which contracts when electricity passes through it. When electricity is gone and once the wire cools down, it will go back to its original length. This type of actuator is not fast, strong, or creates a long stroke. Nonetheless, it is one of the most convenient actuators to use if you need to work with smaller parts.

How to Choose the Proper Actuator for Your Robot

To guide you in choosing the actuator for certain tasks, consider answering the following questions to help you.

Take note that new innovations and technologies are always being released regularly, so nothing is permanent. Also remember that one actuator could perform various tasks in various contexts.

1. Do you need to mobilize a wheeled robot?

Drive motors should carry the weight of the whole robot and will most likely need a gear down. Majority of the robots utilize "skid steering" while trucks or cars utilize rack and pinion steering. If you prefer the skid steering, geared DC motors are recommended to use for robots with tracks or wheels. Geared motors provide constant rotation, and could have discretionary position feedback through optical encoders. Because the rotation needed is limited to a certain angle, you can choose a hobby servo motor for stirring.

2. Is there a limit on the range of motion?

If the range is restricted to 180 degrees and the needed torque is not a critical factor, a hobby servo motor is recommended. Servo motors are available in various torques and sizes and comes with angular position feedback. Majority of these motors use potentiometer, while some specialized ones use optical encoders. R/C servos are now popularly used to build small walking robots.

3. Do you need a motor to lift or turn heavy loads?

Raising a weight needs considerably more power compare to moving a weight on a flat surface. Torque should be prioritized than the speed, and it is ideal to use a gearbox with a powerful DC motor or a linear DC actuator with a high gear ratio. You can use an actuator system that could prevent the mass from falling if there is a disruption in the power source. This includes clamps or worm gears.

4. Do you need the angle to be precise?

Stepper motors that are paired with a motor controller cold provide a very precise angular motion. They are more ideal to use compared to servo motors because they provide constant rotation. But there are also high-end digital servo motors that use optical encoders and can provide high precision.

5. Do you need to achieve movements in a straight line?

Linear actuators are ideal for moving parts and placing them in a straight line. They are available in different configurations and sizes. For fast movements, you must consider solenoids or pneumatics, for high torques, you can use linear DC actuators or hydraulics, and if the movement requires minimum torque, you can use muscle wire.

Chapter 4 - Microcontrollers and Motor Controllers

Microcontrollers are considered as the "brain" of the robot because it is responsible for all decision making, computations, and communications. These are devices with the capacity to execute a program (a series of instructions).

To interact with the external world, a microcontroller has a sequence of electrical signal connections (known as pins), which could be switched on or off using programming functions. These pins are also used in reading electronic signals that are released by sensors or other devices and determine if they are low or high.

Majority of microcontrollers today could measure analogue voltage signals, or signals that could have a full range of values rather than just two specified states by using analog to digital converter or ADC. Through the use of ADC, a microcontroller could assign a numerical value to the analog voltage that is neither low nor high.

What Could Microcontrollers Do?

Numerous complicated actions could be achieved by setting the pins low and high creatively. Nonetheless, building complicated algorithms such as smart movements and data processing or complicated programs are not yet on the range of microcontrollers because of its natural speed and resource limitations.

For example, to light a blinker, you can program a repeating sequence in which the microcontrollers could turn a pin high, wait for several seconds, turn it low, wait for several seconds and goes back to the first sequence. A light that is connected to the pin will then blink open-endedly.

Similarly, microcontrollers could be used to take control of other electronic devices including actuators when they are installed to motor controllers, Bluetooth or WiFi interfaces, storage devices, and many more. Because of its versatility, microcontrollers could be found in common everyday products. Basically, every home electrical device or home appliance utilizes at least one microcontroller.

Not similar to microprocessors found in Central Processing Units in personal computers, microcontrollers don't need peripherals such as external storage devices or external RAM to operate. Hence, even if the microcontrollers are less powerful compared to microprocessors, building circuits and products based on microcontrollers is an easier task and a lot more affordable, because minimal hardware parts are needed. Remember, microcontrollers can output minimum amount of electrical power through pins. Hence, a generic microcontroller cannot power solenoids, power electrical motors, large lights, or other direct loads. Doing this could cause physical damage to the controller.

Programming Microcontrollers

There's no need to shy away from programming microcontrollers. Not similar in the past where making a blinker took comprehensive knowledge of microcontroller and at least a dozen line of code, programming microcontrollers is fairly easy today. You can use the simplified Integrated Development Environments (IDE), which uses modern languages, full line archives that could cover all of the most common actions, and several handy samples to help you get started. You can learn more about programming your robot in Chapter 6.

How to Choose the Proper Microcontroller for Your Robot

You will need a microcontroller for any robotic building project unless you're into BEAM robotics or you want to control your robot through an R/C system or a tether. For starters, selecting the right microcontroller could seem like a difficult job, particularly considering the product range, specifications, and applications. There are various microcontrollers available today such as BasicATOM, POB Technology, Pololu, Arduino, BasicX, and Parallax.

The following questions could guide you in choosing the right microcontroller:

1. Which microcontroller is widely used for your type of robotic project?

Building robots is not a popularity contest, but the fact that a microcontroller has a large supporting community or has been used in the same project can make the design phase easily. With this, you can benefit from other experience and hobbyists. It is common for hobbyists to share codes, pictures, instructions, and videos even lessons learned.

2. Do you need specific accessories for a certain microcontroller?

If your robot has special needs or there is a certain accessory or component that is important for your design, selecting a compatible microcontroller is clearly essential. Even though most accessories and sensors could be directly interfaced with most microcontrollers, some accessories are designed to interface with a particular microcontroller.

3. Do you need special features for your robot?

A microcontroller should be able to perform all the special actions needed for your robot to function well. Some features are common to all microcontrollers such as being able to execute basic mathematical operations, having digital inputs and outputs, and making decisions. Others may need certain hardware such as PWM, ADC, and communication protocol support. You must also consider pin counts, memory and speed requirements

Motor Controllers

Motor controllers are electronic devices that serve an intermediary device between a microcontroller, the motors, and the power supply.

Even though the microcontroller decides the direction and the speed of the motors, it doesn't have enough power to directly drive them. Meanwhile, the motor controller can supply the current at the needed voltage but doesn't have the capacity to decide how fast the motor must turn.

Hence, the microcontroller and the motor controller must work together to make the motors move accordingly. The microcontroller can provide instructions to the motor controller on how to power up the motors through a standard and basic communication method such as PWM and UART. In addition, some motor controllers could be manually regulated using an analog voltage often created through a potentiometer.

The size and weight of a motor controller may greatly vary from a device that is smaller than the tip of a pencil to a huge controller that could weight several kilos. The size and weight often has a minimum effect on the robot, unless you want to build unnamed aerial or aquatic robots.

Types of Motor Controllers

Because there are several types of actuators (as we have discussed in Chapter 3), there are also several types of motor controllers: brushed DC motor controllers, brushless DC motor controllers, servo motor controllers, and stepper motor controllers.

How to Choose a Motor Controller

You can only choose a motor controller after you have decided on what type of actuator you want to use. In addition, the current that a motor draws depends on the torque it could provide. A small DC motor will not use much current, but cannot also release much torque, while a bigger motor could release higher torque but will need increased current.

Chapter 5 - Controlling Your Robot and Use of Sensors

Based on our definition of a robot, it should gather data about its surroundings, make smart decisions and then execute actions based on calculations. This also includes the option for the robot to become semi-autonomous (with aspect that are controlled by humans and other aspects that it can do on its own).

One good example of this is a complex aquatic robot. A human controls the basic motions of the robot while an installed processor measures and reacts to the underwater currents to keep the robot in one position while still preventing a drift. A camera installed in the robot would send videos back to the human while the sensors could track the water pressure, temperature, and more. Once the communication line falters between the robot and the human, an autonomous program could take over to instruct the robot to reach for the surface.

In controlling your robot, you need to figure out the level of autonomy. You need to choose if you want the robot to be tethered, wireless, or autonomous.

Tethered

Direct Wired Control

The simplest way to control a vehicle is by using a handheld controller that is physically connected to a vehicle using a tether or a cable. Knobs, switches, joysticks, buttons, and levers on the controller will allow you to control the robot without the need to add sophisticated electronics. In this setting, the power source and motors could be directly connected with a switch to control the rotation. These machines often have no artificial intelligence and are regarded as remote controlled devices than robots.

Wired Computer Control

Another method is to integrate a microcontroller into the machine but still using a tether. Attaching the microcontroller to your computer's ports will allow you to control the actions using the keyboard, a joystick, a keypad, or other device. Adding a microcontroller to your robot project may also require programming how the robot will respond to the input.

Ethernet

Another way to use computer control is to use an Ethernet interface. A robot that is directly connected to a router can also be used for mobile robots. Building a robot, which can communicate through the internet could be sophisticated, and usually a wireless internet connection is more recommended.

Wireless

Infrared

You can ditch away cables and wires if you use infrared transmitters and receivers. This is often a great achievement for beginners. Infrared control needs "line of sight" to function. The receiver should have the ability to see the transmitter to receive the data. Infrared remote controls can be used to send commands to infrared receivers that are paired with microcontrollers that interpret these signals and control the actions of the robot.

Radio Frequency

Remote control units often use microcontrollers in the receiver and transmitter for data transmission through radio frequency. The receiver box usually has a printed circuit board (PCB) that includes a small servo motor controller and a receiving unit. RF communication needs a transmitter matched with a receiver or a transceiver. RF doesn't need clear line of sight and could also provide considerable distance. Basic RF devices could allow for data transfer between devices between long distances, and there's no limit to the range for more RF devices.

Bluetooth

Bluetooth is a type of Radio Frequency and follows certain protocols for sending and receiving data. Standard Bluetooth range is usually restricted to about 10 meters although it has the advantage of controlling the robot though Bluetooth-enabled devices including laptops, smartphones, and PDAs. Similar to RF, Bluetooth provides two-way communication.

WiFi

Recent development in wireless technology enables you to control a robot through the Internet. To build a WiFi robot, you must have a wireless router that is connected to the internet and a WiFi device on the robot. You can also use a device, which is enabled with TCP/IP with a wireless router.

Autonomous

High-level robots are autonomous. With recent developments, you can now use the microcontroller in its full potential and program it to respond to input from the sensors. Autonomous control may come in different types: restricted sensor feedback, pre-programmed with no feedback from the environment, and complex sensor feedback. Genuine "autonomous" control includes different sensors and code to allow the robot to figure out by itself the smartest action to be taken in any situation.

The most sophisticated methods of control presently used on autonomous robots are auditory and visual commands. For auditory control, a robot will react to the sound of the human's voice for instructions such as "get the ball" or "turn left." For visual command, a robot may look to an object to decide on what to do. Instructing a robot to turn to the right by showing a drawing of an arrow that is pointing to the right requires complicated programming. Even though these things are no longer impossible, they need a sophisticated level programming and usually hundreds of hours.

Not similar to humans, robots are not restricted to just sound, sight, smell, touch, and taste. Robots use different electromechanical sensors to understand and discover their surroundings. Mimicking a natural organism's senses is presently a great challenge, so developers and robotic builders are using alternatives to these natural senses.

Chapter 6 - Assembling and Programming a Robot

After learning all about the fundamental blocks in building a robot, the next stage is the designing and building of the frame that will keep all the components together and will provide your robot a definite look and shape.

Constructing the Frame

There's no fix method in creating a frame, because there is often a trade-off to be constructed. You may prefer to use a lightweight frame but you may need to use costly materials. You may like a strong or big chassis but you may realize it is expensive, hard, or heavy to produce. The frame could be complicated and may take some time to design and build.

Materials

There are different materials that you can use in creating a frame for your robot. As you try different materials to construct not only robots but other types of machines, you will also understand the advantages and the disadvantages as to which material is the most suitable for a specific project. The roster of suggested building materials below comprises only the more common one, and when you have tried several of these materials, you can start experimenting or blending some together.

Basic Construction Materials

Some of the most basic construction materials could be used to build good-quality frames. The cheapest materials is the cardboard that you can usually find for free and could easily be bent, cut, layered, or bent. You can construct a reinforced cardboard box that looks a lot nicer and is more proportional when it comes to the size of your robot. You can then paint it with glue or epoxy to make it stronger then add extra layer of paint.

Structural Flat Materials

For a more durable frame, you can use a standard structural material such as a sheet of plastic, metal, or wood. You just need to puncture some holes to connect the electronic components. A stronger piece of wood has a tendency to be heavy and thick, while a thin sheet of metal could be too flexible. You can attach components to both sides and the wood will still remain solid and intact.

If you're at the stage where you are ready to have an outsourced frame, the best option is to acquire the part precision cut through a water or laser jet. Hiring a third-party to produce a custom part is recommended only if you are completely sure of the dimensions, because the mistakes could be expensive. Companies that offer computer controlled cutting services may also provide different other services such as painting and bending.

3D Printing

Building a frame constructed from 3D printed panels is not always the most structurally sound option, primarily because it is built up in several layers. However, this process could produce complex and detailed shapes that could be impossible to build using other methods. One 3D printed component may contain all the important mounting points for all mechanical or electrical parts without compromising the robot's weight. In the past decade, the cost of 3D printing is quite expensive, but as it becomes popular, the price of producing the components is also expected to go down.

Assembling the Parts Together

With the available options for materials and methods, you can now start assembling the parts together. You can follow the steps below to build a simple, aesthetic, and structurally reliable robot frame.

1. Decide on the material you want to use.

2. Gather all the parts that your robot will need, both mechanical and electrical and measure them. In case you don't have all the components ready, you can refer to the dimensions that are often supplied by the manufacturers.

3. Think of and draw various designs for your frame. It's fine not to provide details.

4. Once you find the suitable design, be certain that the structure is reliable and all the parts would be supported in the frame.

5. Sketch every component of the robot on cardboard or paper at true scale. You can also draw the parts in the CAD software and print them.

6. Test the design in CAD and in actual setting using your paper prototype by test fitting every component and connection.

7. Measure the dimensions again and when you are completely certain that your design is right, begin cutting the frame into the material. Take note to measure two times and only cut once.

8. Test fit every part before assembling the frame if in case you need some changes.

9. Construct the frame using appropriate assembling materials such as glue, nails, screw, duct tape, or any appropriate binding tools that you prefer.

10. Fit all the parts into the frame and there you go, you have just built your robot!

Constructing the Robot Parts

The last step discussed above should be described further. In the past chapters, you have already chosen the electrical parts including the actuators, microcontroller, and motor controller. The next step is to construct them so they will work together.

In the following section, we'll use standard cable colors and terminal names that encompass common parts. You must rely on manuals and datasheets when you are working on your specific parts.

Attaching Motor Controllers to Motors

A geared DC motor or a linear DC actuator usually has two wires: black and red. Attach the black wire to the M- terminal on the DC motor controller and the red wire to the M+ terminal. Connecting the wires the other way around will only cause the motor to rotate in the opposite direction. Meanwhile, servo motors have three wires: red, black, and yellow. A servo motor controller comes with pins that are matching these wires so you can just plug it directly.

Attaching Microcontroller to Motor Controllers

Microcontrollers can communicate with motor controller in different ways: 12C, R/C, Serial, or PWM. Be sure to refer to the manual for each microcontroller for specific instructions on proper connection. Regardless of the method you choose, the microcontroller and the logic of the motor controller should share matching ground reference. This can be achieved by attaching the GND pins together. Meanwhile, a logic level shifter is needed if the devices don't share the same logic levels.

Attaching Batteries to a Microcontroller or a Motor Controller

Majority of the motor controllers available today have two screw terminals for the battery labels marked with B- and B+. If the batteries you got are provided with a connector and the controller comes with screw terminals, you could still search for a pairing connector with wires that you can attach to the screw terminal. If this is possible, you need to find another way to link the battery to the motor controller while you can still unplug the battery and link it to a charger. It's possible that not all the electrical and mechanical components you have selected for your robot could operate on a single voltage, and so may need several voltage regulation circuits or batteries.

If you are building a robot with a microcontroller, DC gear motors, and maybe servo motors, it's easy to see how a battery may not be able to power every component directly. Nevertheless, it is best to choose a battery that can directly power as many devices that you need. The battery with the largest capacity must be connected with the drive motors. For instance, if the motors you select are rated a nominal 12 volts, the primary battery must also be 12 volts. So you can use a regulator to energize a 5 volts microcontroller. LiPo and NiMH batteries are the top choices for small to medium robots. Select NiMH if you need cheaper batteries and LiPo if you need light weight batteries. Always take note that batteries are powerful devices that could easily burn your circuits if they are not properly connected. Always make sure that the polarity is correct and that your device could handle all the energy supplied by the battery. If you are not certain, never make assumptions.

Adding Electrical Parts to Frame

You can attach electrical components to your frame through different methods. Make certain that whatever method you use, don't conduct electricity. Usual methods include screws, hex spacers, Velcro, double-sided tape, cable ties, glue, and many more.

Programming Your Robot

Programming is often the last step in building your robot. If you have followed the steps described in the previous chapters, by now you have selected the electrical components such as actuators, microcontrollers, motor controllers, sensors, and more. At this point, you might have already constructed your robots and hopefully it looks something like you want it to be. But without the proper program, your robot is just a cool paperweight.

It requires another book to teach you robotic programming. Instead, this section will guide you on how to get started and what you should learn.

There are several programming languages that you can use to program the microcontroller that will serve as the brain of your robot. The following are the most common programming languages you can choose:

Assembly

This programming language is just a shy away from programming a full-pledged computer, and so it could be difficult to use. This language is ideal to use if you really need to ensure complete instruction-level control of your robot.

Basic

Basic is one of the most common programming languages for robot hobbyists. This is often used in programming microcontrollers mainly for educational robots.

C++

C++ is a very popular programming language. It provides top=level functionality while you are keeping a good low-level control. A variant of C++ is Processing, which includes simplified codes to make the programming easier.

Java

Java is more developed compared to C++ and offers any safety features to the disadvantage of low-level control. Some producers of microcontrollers such as Parallax are making components for specific use with Java.

Python

Python is one of the most popular languages for scripting. It is easy to learn and could be used to quickly and efficiently integrate several programs.

If you have selected a hobbyist type of microcontroller from a known producer, there's a chance that you can find a book that you can read so you can learn how to program in their preferred programming language. But if you instead prefer a microcontroller from a smaller producer, it is crucial to see what language the controller wants to use and what tools are available.

Conclusion

Thank you again for purchasing this book!

I hope this book was able to help you to learn the basic building blocks of robot building.

The next step is to expand your knowledge in robotics, especially learning advanced programming for your robot.

Finally, if you enjoyed this book, please take the time to share your thoughts and post a review on Amazon. It'd be greatly appreciated!

Thank you and good luck!

www.ingramcontent.com/pod-product-compliance
Lightning Source LLC
Chambersburg PA
CBHW071827200526
45169CB00018B/1156